VOLUME FIFTY TWO

THE ENZYMES

DNA Damage and Double Strand Breaks - Part B

Serial Editors

LAURIE S. KAGUNI

Department of Biochemistry and Molecular Biology, Michigan State University, East Lansing, Michigan, United States

FUYUHIKO TAMANOI

University of California, Los Angeles, USA

VOLUME FIFTY TWO

THE ENZYMES

DNA Damage and Double Strand Breaks - Part B

Edited by

FUYUHIKO TAMANOI
Institute for Integrated Cell-Material Sciences, Institute for Advanced Study, Kyoto University, Kyoto, Japan

KENICHI YOSHIKAWA
Center for Integrative Medicine and Physics, Kyoto University; Faculty of Life and Medical Sciences, Doshisha University, Kyoto, Japan

ELSEVIER

ACADEMIC PRESS
An imprint of Elsevier

Academic Press is an imprint of Elsevier
50 Hampshire Street, 5th Floor, Cambridge, MA 02139, United States
525 B Street, Suite 1650, San Diego, CA 92101, United States
The Boulevard, Langford Lane, Kidlington, Oxford OX5 1GB, United Kingdom
125 London Wall, London, EC2Y 5AS, United Kingdom

First edition 2022

ISBN: 978-0-323-99398-2
ISSN: 1874-6047

For information on all Academic Press publications
visit our website at https://www.elsevier.com/books-and-journals

Publisher: Zoe Kruze
Acquisitions Editor: Leticia Lima
Developmental Editor: Leticia Lima
Production Project Manager: Abdulla Sait
Cover Designer: Christian Bilbow

Typeset by STRAIVE, India

Contents

Contributors

Fuyuhiko Tamanoi
Institute for Integrated Cell-Material Sciences, Institute for Advanced Study, Kyoto University, Kyoto, Japan; Dept. of Microbiology, Immunology & Molecular Genetics, University of California, Los Angeles, CA, United States

Kenichi Yoshikawa
Center for Integrative Medicine and Physics, Kyoto University; Faculty of Life and Medical Sciences, Doshisha University, Kyoto, Japan

Preface

DNA damage including base modifications and DNA breaks is of major significance for cell physiology. If not repaired, these lesions will lead to mutations and cell death. Of particular importance is double-strand breaks that are difficult to repair. Volume 51 dealt with the topic of DNA double-strand breaks. In this volume, we further expand the discussion by providing a summary of Volume 51 and then discussing future issues.

The history of the *Enzymes* series goes back to 1950 and a large number of topics including DNA enzymes and biochemistry have been published. We recently started a project to document these old works. In this volume, we have included chapters that describe the contents of volumes that were published as the First Edition and the Second Edition. We hope this will provide information on how the *Enzymes* series has evolved.

Fuyuhiko Tamanoi
Kenichi Yoshikawa
September 3, 2022
Kyoto, Japan

CHAPTER ONE

Summary of volume 51: DNA damage and double-strand breaks

Fuyuhiko Tamanoi[a,*] and Kenichi Yoshikawa[b,c]
[a]Institute for Integrated Cell-Material Sciences, Institute for Advanced Study, Kyoto University, Kyoto, Japan
[b]Faculty of Life and Medical Sciences, Doshisha University, Kyoto, Japan
[c]Center for Integrative Medicine and Physics, Kyoto University, Kyoto, Japan
*Corresponding author: e-mail address: tamanoi.fuyuhiko.2c@kyoto-u.ac.jp

Contents

Abstract

DNA damage and breaks are events that happen to DNA which exert a variety of influence on cell physiology including inhibition of DNA synthesis, repair response, cell cycle effect and cell death. Thus, it is important to deepened understanding of these events. In volume 51, we discussed topics including (1) assays to detect double-strand breaks, (2) conditions leading to double-strand breaks, (3) effects of irradiation, (4) DNA structure and chromatins, and (5) direct and indirect effect on DNA. Contributing authors and a table of contents for volume 51 are mentioned. We also discuss further issues and topics that need to be featured in future volumes. These include DNA damage sensors, DNA damage response proteins, and double-strand break repair pathways.

Abbreviations

BNCT boron neutron capture therapy
DSB double-strand break
H2A histone 2A
TUNEL terminal deoxynucleotidyl transferase dUTP nick end labeling

The Enzymes, Volume 52
ISSN 1874-6047
https://doi.org/10.1016/bs.enz.2022.10.001

1. Summary of volume 51 (DNA damage and double-strand breaks, Part A)

1.1 Assays for detecting double-strand breaks

DNA is under continuous assault and, as a result, suffers from DNA damage and breaks. We focus on DNA breaks that include single-strand and double-strand breaks. Double-strand breaks are more serious, as they are more difficult to repair and the failure to repair these DNA damages lead to inhibition of DNA replication leading to cell death.

A variety of methods have been developed to detect DNA double-strand breaks. The most frequently used methods employ phosphorylation of histone H2A protein and the method is called "γH2AX assay." This is discussed in "DNA damage and biological responses induced by boron neutron capture therapy (BNCT)" by Natsuko Kondo; "Auger electrons and DNA double-strand breaks studied by using iodine-containing chemicals" by Yuya Higashi, Yue Ma, Kotaro Matsumoto, Ayumi Shiro, Hiroyuki Saitoh, Tetsuya Kawachi and Fuyuhiko Tamanoi; and "Carbon ion radiation and clustered DNA double-strand breaks" by Atsushi Shibata. This method uses an antibody specific for the phosphorylated form of histone H2A. Upon double-strand break formation, histone H2A is phosphorylated and this provides a marker for the DSBs. After fixing cells, γH2AX-positive cells can be detected. This method can be used in a single cell situation thus providing significant details regarding the DNA breaks.

Single molecule based analyses of double-strand break are discussed in "Quantitative evaluation of DNA double-strand breaks (DSBs) through single-molecule observation" by Kenichi Yoshikawa. The assay uses DNA of a defined length such as lambda phage DNA. Double-strand breaks lead to cleavage of double-strand DNA that could be detected by electron microscopy. This method provides quantitative and temporal assessment of DNA breaks and contributed to significant increase in our understanding of DNA breaks. "Mechanical force induced DNA double-strand breaks: Ultrasound" by Yue Ma and Iwaki Akiyama uses this assay to examine effect of ultrasound on the formation of double-strand breaks. "Damages of DNA in tritiated water" by Yuji Hatano, Hiroaki Nakamura, Susumu Fujiwara, Seiki Saito, Takahiro Kenmotsu employs this method to analyze effect of β-radiation by tritium water.

1.2 Conditions leading to double-strand break formation

Double-strand breaks are generated by internal events as well as external events. Internal events including metabolic changes result in the generation of reactive oxygen species which then cause DNA damage. External events include exposure to various insults such as environmental carcinogen exposure as well as exposure to UV, X-ray and others.

An important type of insult that results in double-strand breaks concerns mechanical stress. "Mechanical force induced DNA double-strand breaks: Ultrasound" by Yue Ma and Iwaki Akiyama discuss one such example, effect of ultrasound on DNA double-strand breaks. Intensity and duration of ultrasound treatment were systematically altered thus revealing features of DNA breaks in response to ultrasound. Ultrasound is a promising treatment for cancer therapy and the study has significant implication for therapies.

1.3 Particles, clustered breaks, radiation

Radiation is widely used for cancer therapy. The most frequently used method involves X-ray where photons deposit energy to DNA resulting in DNA breaks. Photons also interact with molecules in the cell such as water molecules generating reactive oxygens when then damage DNA. Instead of photons, proton beams and heavy ion beams such as carbon beams are also used. Irradiation with protons and carbon ions has the advantage to deposit energy at the site of tumor instead of at the surface of a patient, due to difference in the Bragg peak. Boron neutron capture therapy (BNCT) uses neutron beam that is irradiated onto boron-10. This results in the splitting of boron to yield lithium and helium nucleus. The helium nucleus is an α-particle that has strong effect on DNA. Another type of particles that influence DNA are Auger electrons generated by the irradiation of high Z elements such as gold, silver, gadolinium or iodine with monochromatic X-rays. The Auger electrons travel only a short distance but produce strong DNA damaging effect resulting in double-strand breaks. The Auger electrons and carbon ions cause complex DNA breaks. "Boron neutron capture therapy (BNCT) and DNA breaks" by Natsuko Kondo discusses BNCT, while "Carbon ion radiation and clustered DNA double-strand breaks" by Atsushi Shibata discusses carbon ion radiation. "Core level ionization or excitation and Auger relaxation induces clustered DNA damage" by Akinari Yokoya and Yui Obata and "Auger electrons and DNA double-strand breaks studied by using iodine-containing chemicals" by Yuya Higashi, Yue Ma, Kotaro Matsumoto, Ayumi Shiro, Hiroyuki Saitoh, Tetsuya Kawachi and Fuyuhiko Tamanoi discuss Auger electrons. β-particles are discussed in "Damages of DNA in tritiated water" by Yuji Hatano, Hiroaki Nakamura, Susumu Fujiwara, Seiki Saito, and Takahiro Kenmotsu.

1.4 DNA structure and chromatins

Genomic DNAs in living cells exhibit the size on the order of Mbp—100Mbp, corresponding to the length scale of the order of mm—10 cm. On the other hand, the diameter of double-stranded DNA is around 2 nm. It is known that double-stranded DNA is characterized as rather stiff double-helical structure, where the persistence length of double-stranded DNA is around 50 nm (corresponding to 150–200 bp), which indicates that long genomic DNAs behaves as semi-flexible polymer chain, whereas short oligomeric DNAs exhibit the characteristics as stiff rods. As DSBs cause serious effect on living cells even with one break for long genome-sized DNA molecules, it is critically important to perform the measurement and evaluation on the generation of DSBs for giant DNAs, by taking into account of the effect of their higher-order structure. As has been explained in "Quantitative evaluation of DNA double-strand breaks (DSBs) through single-molecule observation" by Kenichi Yoshikawa, long DNAs above the size of several tens kbp perform folding transition between elongated coil state and folded compact state [1,2], accompanied by the change of the solution environment. On the other hand, short oligomeric DNAs exhibit aggregation without the generation of the folding transition. "Quantitative evaluation of DNA double-strand breaks (DSBs) through single-molecule observation" by Kenichi Yoshikawa describes several experimental observations on significant protective effect against DSBs for the genomic DNAs under folded compact states, based on the methodology of single DNA observation by fluorescence microscopy.

In eukaryotes, genomic DNAs associate with positively charged histone proteins to form chromatin. In its elemental structure, each nucleosome is formed through the wrapping of DNA with a core histone octamer. Nucleosome particles are arranged assisted by linker histones as bead-on-a-string structure. For the larger scale, such beads-on-a-string is folded into condensed state in chromatin. As explained in "Chromatin organization and DNA damage" by Katsuhiko Minami, Shiori Iida, and Kazuhiro Maeshima, recent 3C/Hi-C methods have uncovered the hierarchy organization of the genome. It was revealed that the genome in higher eukaryotes is subdivided into chromatin domains; topologically associating domains (TADs) and smaller loop domains on the order of 200 kbp. These chromatin domains can be further bundled into mega-base scaled compartments. Thus, in interface nuclei, chromatin is composed with numerous compact domains as functional units of the genome, and 3D structural arrangement of the domains neatly concerns with the fundamental mechanism of regulation with a variety of genome functions, including epigenetics [3].

In "Chromatin organization and DNA damage" by Katsuhiko Minami, Shiori Iida, and Kazuhiro Maeshima, experimental studies with direct observation of DSBs in chromatin by applying TUNEL assay (with terminal deoxynucleotidyl transferase, TdT) [4] are described. Among many novel observations, one of the most important discovery would be the general trend that the protective effect on condensed chromatin is derived from chromatin compaction itself, but not from the level of chromatin-associated proteins. Previous work implied that chromatin-associated proteins (non-histone proteins) could protect against DNA damage [5–7], where higher protein levels had a stronger ability to prevent chromatin damage caused by ionizing irradiation. In contrast, it has been confirmed that associated proteins did not protect chromatin from damage as decondensed and recondensed chromatin possessed similar total amounts and types of associated chromatin proteins.

1.5 Direct and indirect effect

Currently, DSBs induced by ionization radiation has been extensively studied by *in vitro* experiments and cell imaging techniques such as TUNEL assay [4]. Usually in these studies, the damages have been discussed in terms of the direct and indirect effect of radiation. The direct effect is due to ionization or electronic excitation of the DNA caused by the radiation source. The indirect effect is caused by reactive oxygen species, in most cases free radicals, which are produced in the vicinity of the DNA. In spite of the hard efforts to shed light onto the mechanism of DSBs in terms of direct and indirect effects, it has been rather difficult to determine the actual chemical mechanism because individual DNA damages are induced on the scale of 1 nm or less. In "Quantitative evaluation of DNA double-strand breaks (DSBs) through single-molecule observation" by Kenichi Yoshikawa and "Mechanical force induced DNA double-strand breaks: Ultrasound" by

Yue Ma and Iwaki Akiyama, it is shown that single molecular observation on DNA molecules provides useful insights on the kinetics of DSBs; that is, whether the breakage reaction is caused by single-step or sequential steps. By denoting that x is the number of DNA molecules surviving from fragmentation without DSB, the reaction rate, dx/dt, is given as follows [8–10]:

For the single-step DSB:

$$\ln x \sim -t$$

For the DSB induced by two-step reaction with twin single-strand breaks:

$$\ln x \sim -t^2$$

Thus, when double-strand breakage is induced in a single-step, a plot of $\ln x$ vs. t should be a straight line. On the other hand, in a two-step reaction with stochastic single-strand damage, a plot of $\ln x$ vs. t^2 should provide a straight line. It is highly expected that the so-called direct effect would correspond to single-step DSB and the indirect effect may induce two-step reaction kinetics. It is noted that such analysis based on the quantitative analysis of the reaction kinetics is useful for the quantitative evaluation of the protective effects of various anti-oxidants.

2. Alphabetical list of authors and chapter contents

Akiyama, Iwaki
Faculty of Life and Medical Science, Doshisha University, Kyotanabe, Japan
Fujiwara, Susumu
Faculty of Materials Science and Engineering, Kyoto Institute of Technology, Kyoto 606-8585, Japan
Hatano, Yuji
Hydrogen Isotope Research Center, University of Toyama, Toyama, 930-8555, Japan
Higashi, Yuya
Institute for Integrated Cell-Material Sciences, Institute for Advanced Study, Kyoto University, Kyoto, Japan
Iida, Shiori
Genome Dynamics Laboratory, National Institute of Genetics, Research Organization of Information and Systems (ROIS), Mishima, Shizuoka 411-8540, Japan; Department of Genetics, School of Life Science,

SOKENDAI (Graduate University for Advanced Studies), Mishima, Shizuoka 411-8540, Japan
Kawachi, Tetsuya
Kansai Photon Science Institute, Quantum Beam Science Research Directorate, National Institutes for Quantum Science and Technology, Hyogo, Japan
Kenmotsu, Takahiro
Faculty of Life and Medical Sciences, Doshisha University, Kyoto 610-0321, Japan
Kondo, Natsuko
Particle Radiation Oncology Research Center, Institute for Integrated Radiation and Nuclear Science, Kyoto University, Osaka, Japan
Ma, Yue
Institute of Biomedical Engineering, Chinese Academy of Medical Sciences and Peking Union Medical College, Tianjin, China
Maeshima, Kazuhiro
Genome Dynamics Laboratory, National Institute of Genetics, Research Organization of Information and Systems (ROIS), Mishima, Shizuoka 411-8540, Japan; Department of Genetics, School of Life Science, SOKENDAI (Graduate University for Advanced Studies), Mishima, Shizuoka 411-8540, Japan
Matsumoto, Kotaro
Institute for Integrated Cell-Material Sciences, Institute for Advanced Study, Kyoto University, Kyoto, Japan
Minami, Katsuhiko
Genome Dynamics Laboratory, National Institute of Genetics, Research Organization of Information and Systems (ROIS), Mishima, Shizuoka 411-8540, Japan; Department of Genetics, School of Life Science, SOKENDAI (Graduate University for Advanced Studies), Mishima, Shizuoka 411-8540, Japan
Nakamura, Hiroaki
National Institute for Fusion Science, Toki City, Gifu 509-5292, Japan
Obata, Yvui
Institute for Quantum Life Science, QST, 4-9-1 Anagawa, Inage-ku, Chiba-shi 263-8555, Japan; Graduate School of Science and Engineering, Ibaraki University, Mito, Ibaraki 310-8512, Japan
Saito, Hiroyuki
Kansai Photon Science Institute, Quantum Beam Science Research Directorate, National Institutes for Quantum Science and Technology, Hyogo, Japan

Shibata, Atsushi
Gunma University Initiative for Advanced Research, Gunma University, Maebashi, Gunma, Japan
Saito, Seiki
Graduate School of Science and Engineering, Yamagata University, Yamagata, Japan
Shiro, Ayumi
Kansai Photon Science Institute, Quantum Beam Science Research Directorate, National Institutes for Quantum Science and Technology, Hyogo, Japan
Tamanoi, Fuyuhiko
Institute for Integrated Cell-Material Sciences, Institute for Advanced Study, Kyoto University, Kyoto, Japan
Yokoya, Akinari
Institute for Quantum Life Science, QST, 4-9-1 Anagawa, Inage-ku, Chiba-shi 263-8555, Japan; Graduate School of Science and Engineering, Ibaraki University, Mito, Ibaraki 310-8512, Japan
Yoshikawa, Kenichi
Faculty of Life and Medical Science, Doshisha University, Kyotanabe, Japan; Institute for Advanced Study, Kyoto University, Kyoto

CONTENTS

Chapter 7: Auger electrons and DNA double-strand breaks studied by using iodine-containing chemicals
Yuya Higashi, Yue Ma, Kotaro Matsumoto, Ayumi Shiro, Hiroyuki Saitoh, Tetsuya Kawachi and Fuyuhiko Tamanoi
Chapter 8: Carbon ion radiation and clustered DNA double-strand breaks
Atsushi Shibata
Chapter 9: Damages of DNA in tritiated water
Yuji Hatano, Hiroaki Nakamura, Susumu Fujiwara, Seiki Saito, Takahiro Kenmotsu

3. Further topics to be discussed in future volumes

In future volumes, we plan to address the question what happens after DNA double-strand breaks are generated inside the cell. Cells have DNA damage sensors that detect double-strand breaks and recruit DNA damage response proteins. If we take DNA damage induced by ionizing radiation for example, there are three events that need to be considered [11]. One is DNA damage sensors (DDS). The second is the recruitment of DNA damage response (DDR) proteins. Finally, DNA double-strand break repair occurs and this involves DSB repair pathways.

When double-strand breaks are generated, H2AX, a variant of the core histone protein H2A is phosphorylated at the S139 site [12]. This results in the formation of γH2AX foci that can be visualized by the use of a monoclonal antibody against phosphorylated γH2AX. DNA repair proteins such as BRCA1 (BRCA1 C terminal), p53-binding protein 1 (53BP1), MDC1 and RAD1 are recruited to the γH2AX foci.

The MRN complex that consist of Nbs1, hMre11, and hRAD50 links DSB repair with cell cycle checkpoint functions [13]. The Nbs1/hMre11/hRad50 complex plays an important role in the signal transduction initiated by X-ray radiation [14]. One of the roles of this complex is to recruit activated ATM (ataxia-telangiectasia mutated) to the site of DNA damage.

Primary repair pathways for double-strand breaks involve two classical pathways, NHEJ (nonhomologous end-joining) and homologous recombination (HR) pathways. The HR pathway was discovered first and was shown to be triggered in cells in the late S and G2/M phases [15]. The second repair pathway NHEJ was discovered in the 1980s and this pathway is triggered in the G0/G1 phase as well as G2/M phase. In mammalian cells, the NHEJ pathway is a predominant pathway [16,17].

References

[1] K. Yoshikawa, Y. Yoshikawa, Compaction and condensation of DNA, in: R.I. Mahato, S.W. Kim (Eds.), Pharmaceutical Perspectives of Nucleic Acid-Based Therapy, Taylor & Francis, London, 2002, pp. 136–163.

[2] A. Estevez-Torres, D. Baigl, DNA compaction: fundamentals and applications, Soft Matter 7 (2011) 6746–6756.

[3] K. Maeshima, A phase transition for chromosome transmission when cells divide, Nature 609 (2022) 35–36.

[4] Y. Gavrieli, Y. Sherman, S.A. Ben-Sasson, Identification of programmed cell death in situ via specific labeling of nuclear DNA fragmentation, J. Cell Biol. 119 (1992) 493–501.

[5] R.L. Warters, B.W. Lyons, Variation in radiation-induced formation of DNA double-strand breaks as a function of chromatin structure, Radiat. Res. 130 (1992) 309–318.

[6] M.C. Elia, M.O. Bradley, Influence of chromatin structure on the induction of DNA double strand breaks by ionizing radiation, Cancer Res. 52 (1992) 1580–1586.

[7] J. Nygren, M. Ljungman, G. Ahnstrom, Chromatin structure and radiation-induced DNA strand breaks in human cells: soluble scavengers and DNA-bound proteins offer a better protection against single- than double-strand breaks, Int. J. Radiat. Biol. 68 (1995) 11–18.

[8] Y. Yoshikawa, T. Mori, N. Magome, K. Hibino, K. Yoshikawa, DNA compaction plays a key role in radioprotection against double-strand breaks as revealed by single-molecule observation, Chem. Phys. Lett. 456 (2008) 80–83.

[9] Y. Yoshikawa, T. Mori, M. Suzuki, T. Imanaka, K. Yoshikawa, Comparative study of kinetics on DNA double-strand break induced by photo and gamma-irradiation: Protective effect of water-soluble flavonoids, Chem. Phys. Lett. 501 (2010) 146–151.

[10] M. Noda, Y. Ma, Y. Yoshikawa, T. Imanaka, T. Mori, M. Furuta, T. Tsuruyama, K. Yoshikawa, A single-molecule assessment of the protective effect of DMSO against DNA double-strand breaks induced by photo-and γ-ray-irradiation, and freezing, Sci. Rep. 7 (8557) (2017) 1–8.

[11] R.-X. Huang, P.-K. Zhou, DNA damage response signaling pathways and targets for radiotherapy sensitization in cancer, Signal Trans. Targeted Therapy 5 (2020) 1–27.

[12] M.S. Siddiqui, M. Francois, M.F. Fenech, W.R. Leifert, Mutat. Res. Rev. Mutat. Res. 766 (2015) 1–19.

[13] J.P. Carney, R.S. Maser, H. Olivares, E.M. Davis, M. Le Beau, J.R. Yates III, L. Hays, W.F. Morgan, J.H.J. Petrini, The hMre11/hRad50 protein complex and Nijmegen breakage syndrome: linkage of double-strand break repair to the cellular DNA damage response, Cell 93 (1998) 477–486.

[14] Y. Habraken, O. Jolois, J. Piette, Differential involvement of the hMRE11/hRAD50/NBS1 complex, BRCA1 and MLH1 in NF-κB activation by camptothecin and X-ray, Oncogene 22 (2003) 6090–6099.

[15] M.A. Resnick, The repair of double-strand breaks in DNA: a model involving recombination, J. Theor. Biol. 59 (1976) 97–106.

[16] D.B. Roth, J.H. Wilson, Relative rates of homologous and nonhomologous recombination in transfected DNA, Proc. Natl. Acad. Sci. U. S. A. 82 (1985) 3355–3359.

[17] D.B. Roth, T.N. Porter, J.H. Wilson, Mechanisms of nonhomologous recombination in mammalian cells, Mol. Cell. Biol. 5 (1985) 2599–2607.

CHAPTER TWO

Inhibition of DNA synthesis and cancer therapies

Fuyuhiko Tamanoi[a,*] and Kenichi Yoshikawa[b,c]
[a]Institute for Integrated Cell-Material Sciences, Institute for Advanced Study, Kyoto University, Kyoto, Japan
[b]Faculty of Life and Medical Sciences, Doshisha University, Kyoto, Japan
[c]Center for Integrative Medicine and Physics, Kyoto University, Kyoto, Japan
*Corresponding author: e-mail address: tamanoi.fuyuhiko.2c@kyoto-u.ac.jp

Contents

Abstract

Cancer is a worldwide problem afflicting 19 million people. Inhibition of DNA synthesis has been a cornerstone of anticancer therapy. A variety of chemotherapy drugs have been developed and many of these are aimed at inhibiting DNA synthesis, as they damage DNA, form DNA adduct and interfere with DNA synthesis. Another type of chemotherapy interferes with the synthesis of nucleotide pools. There are also other types of drugs that inhibit topoisomerases resulting in the interference with DNA replication and transcription. Significant progress has been made regarding radiation therapy that includes X-ray (and γ-ray), proton therapy and heavy ion therapy. The Auger therapy is a type of radiation therapy that differs from X-ray, proton or heavy ion therapy. The method relies on the use of high Z elements such as gadolinium, iodine, gold or silver. Irradiation of these elements results in the release of electrons including the Auger electrons that have strong DNA damaging effect. Tamanoi *et al.* developed novel nanoparticles containing gadolinium or iodine to place high Z elements at the periphery of the nucleus thus localizing them close to DNA. Irradiation with monochromatic

The Enzymes, Volume 52
ISSN 1874-6047
https://doi.org/10.1016/bs.enz.2022.10.002

X-ray resulted in the formation of double-strand DNA breaks leading to the destruction of tumor mass. Comparison of conventional X-ray therapy and the Auger therapy is discussed.

1. Cancer

In 2021, the number of cancer patients worldwide is reported to be 19 million and the number of cancer death is 10 million [1]. Thus, cancer is a worldwide problem and the number of cancer death is staggering. While there has been a significant progress in cancer therapy, treatment options remain limited for some types of cancer. For example, pancreatic cancer remains a major problem and no effective treatment options for advanced stage patients exist [2]. Ovarian cancer is also a type of cancer that is difficult to treat [3]. In addition, scirrhous gastric cancer is a type of stomach cancer that is associated with rapid progression of the disease and lack of effective treatment [4]. Recurrence of cancer is a major problem. Even when cancer responds to the initial chemotherapy treatment, patients experience cancer recurrence almost all the cases. When this happens, the recurrent tumor becomes resistant to chemotherapy. Thus, it is important to continue to develop new types of cancer treatments.

2. Chemotherapy

A variety of anticancer drugs have been developed and used in clinics. A number of these drugs target DNA synthesis as described in Fig. 1. Among them, DNA-targeting cytotoxic drugs have been frequently used for medical treatments. They can be grouped based on their mode of action; alkylating agents, platinum-based materials, antimetabolites, topoisomerase inhibitors and antitumor antibiotics.

2.1 DNA adducts forming agents

Alkylating agents act by forming DNA adducts, as these agents react directly with the oxygen and nitrogen atoms of DNA bases [5]. These agents include temozolomide and dacarbazine that form an adduct on one of the DNA strands, as well as aziridines and epoxides that can react with two strands resulting in the formation of inter-strand crosslinks (ICLs) [6]. Cisplatin, carboplatin and oxaliplatin are platinum-based compounds that react with DNA. These compounds are hydrolyzed after entering the cell

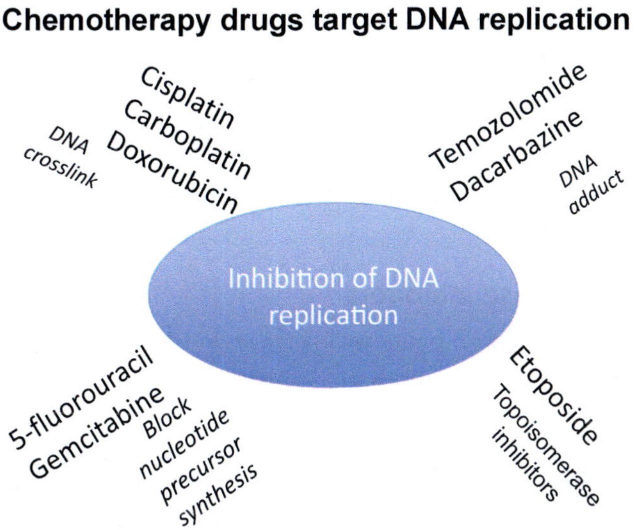

Fig. 1 Chemotherapy drugs target DNA replication. Drugs such as cisplatin, carboplatin and doxorubicin form DNA crosslinks while drugs such as temozolomide and dacarbazine form DNA adducts. Drugs such as 5-fluorouracil and gemcitabine block nucleotide precursor synthesis. In addition, topoisomerase inhibitors are used in chemotherapy.

resulting in the formation of platinum-containing moiety with two DNA reactive sites. This enables formation of crosslinks (both inter and intra-strand) [7–9].

2.2 Agents that interfere with DNA metabolism

Antimetabolites represent another class of chemotherapy drug that mimics molecules that are essential for DNA replication. They can deplete deoxynucleoside triphosphate (dNTP) needed for DNA synthesis. One example is 5-fluorouracil (5-FU) that can inhibit the synthesis of thymidine. In addition, 5-FU can be incorporated into DNA leading to the inhibition of DNA synthesis [10,11]. Gemcitabine is a cytidine analog mentioned in Fig. 1. The 2′ position of the ribose of cytidine is fluoride in this analog [12]. It is incorporated into DNA and terminate DNA synthesis. This drug is used in a variety of therapies including pancreatic cancer therapy; this drug is used as one of the first line treatments for pancreatic cancer. In clinical settings, gemcitabine is used in combination with abraxane, BSA-coated paclitaxel drug. This chemical appears to exhibit less side effect compared to platin-based drugs.

Antitumor antibiotics include anthracyclines, mitomycin C and bleomycin. Doxorubicin is an example of anthracyclines that intercalates in the DNA and functions as a topoisomerase poison [13,14]. Mitomycin C forms covalent linkages with DNA and functions as an alkylating agent [15]. Bleomycins are classified as "radiomimetic", as they mimic ionizing radiation effect by the generation of ROS (reactive oxygen species) [16].

2.3 Topoisomerase inhibitors

During DNA replication, overwinding or underwinding of DNA occurs. Enzymes that regulate this event are topoisomerases [17]. Chemicals that stabilizes the topoisomerase-DNA complex are called "topoisomerase poisons" [18]. Chemicals that form topoisomerase poisons with type I topoisomerase include topotecan and irinotecan. On the other hand, chemicals such as etoposide form a poison with type II topoisomerase.

3. Radiation therapy

3.1 Various types of radiation therapy

Radiation causes DNA damage such as DNA modifications, generation of single-strand breaks as well as double-strand breaks. This leads to the inhibition of DNA synthesis and induction of apoptosis. Conventional radiation therapy uses X-ray or γ-ray and is widely used with efficacy for a variety of cancer (Fig. 2A). In the US, 60% of cancer patients receive radiation therapy. However, this approach suffers from energy deposition at the skin before reaching the site of tumor. This is because the energy deposition occurs highest at the surface and decreases as the beam penetrates into the body. Thus, skin lesions occur. To prevent this problem, radiation from multiple directions focusing on the tumor is pursued. There has been remarkable progress with the X-ray radiation therapy [19].

More recently, use of ionizing beams such as protons and heavy ions [20,21] has been developed (Fig. 2B and C). Carbon ion beam has been used for heavy ion source. In contrast to photons, proton and carbon ion beams have a different energy deposition pattern. Because of their Bragg peak feature, it is possible to deposit majority of beam energy to the tumor. Proton centers and carbon ion centers have been developed all over the world.

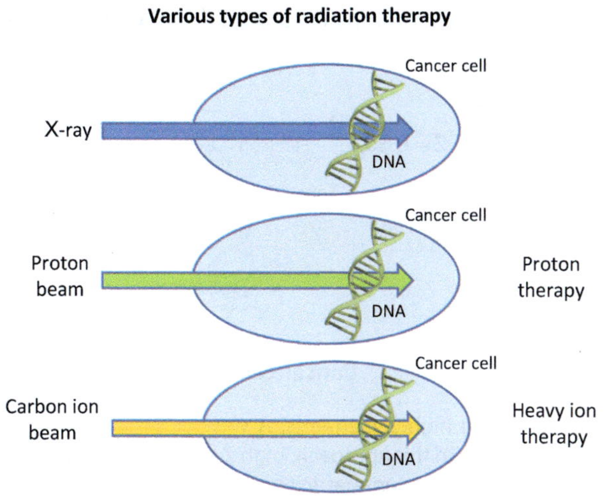

Fig. 2 Various types of radiation therapy. In addition to the conventional radiation therapy that uses X-ray or γ-ray, proton therapy and heavy ion therapy have been developed.

3.2 Auger therapy and a quest for a new type of radiation therapy

The Auger therapy is based on the observations made by Lisa Meitner and Pierre Auger [22] that when high Z atoms (such as gadolinium, iodine, gold or silver) is irradiated with monochromatic X-ray having an energy that is close to the energy of K-shell electrons, then the X-ray energy is absorbed by the K-shell electron. This causes destabilization of the atom and an electron from a higher shell such as L-shell moves down releasing an energy. This energy is absorbed by an electron leading to the release of multiple electrons that are called the Auger electrons. Thus, the irradiation with monochromatic X-ray results in the release of photoelectrons and the Auger electrons.

Special features of the Auger electrons are noteworthy. First, they have a short flight distance in the order of nanometer. Second, they have strong DNA damaging effect causing not only double-strand DNA breaks but also clustered damages which are difficult to repair. There have been a number of attempts to employ the Auger effect as a means for cancer therapy [23]. Our approach is to place high Z atoms close to the nucleus where DNA resides (Fig. 3).

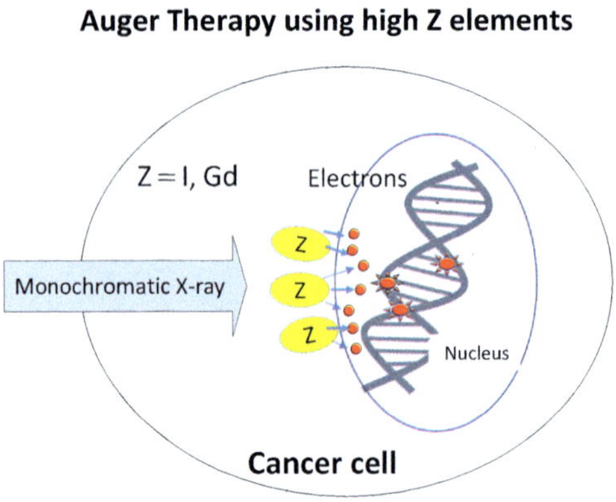

Fig. 3 Auger therapy using high Z elements. A view of the Auger therapy that uses high Z element is presented. In this approach, high Z elements are placed at the periphery of cell nucleus by the use of nanoparticles.

3.3 Kyoto university experiments employing nanoparticles and tumor spheroids

Tamanoi and his colleagues have recently developed nanoparticles containing gadolinium or iodine (Fig. 4) and used a spheroid-based assay to demonstrate the efficacy to destroy tumor mass [24,25]. They used mesoporous silica nanoparticles (MSN) which are 100–200 nm size particles prepared by the sol-gel synthesis. These nanoparticles contain thousands of pores and look like Swiss-cheese or honey comb. Because the pores also contain surface area, a key feature of this type of nanomaterials is the presence of a large surface area. In fact, the surface area of 1 g of MSN is calculate to be about 100 m^2. To attach gadolinium, MSN surface was modified with amine groups which enabled interaction with gadopentetic acid (these are called Gd-MSN). For the preparation of iodine containing nanoparticles, biodegradable nanoparticles containing tetrasulfide bonds in the framework were prepared (these are called IPO for iodine containing porous organosilica). Towards the end of the synthesis, iodine containing precursor was added so that the nanoparticles contain iodine close to the surface.

Both Gd-MSN and IPO were efficiently taken up into human cancer cells and were localized at the periphery of cell nucleus (Fig. 5). This enabled localization of high Z elements in a concentrated fashion close to DNA. In addition, they were efficiently taken up into tumor spheroids prepared from human ovarian cancer cells. The tumor spheroids are usually 0.3 mm diameter and contain 10^5 cells. Importantly, nanoparticles were distributed evenly within the spheroids. This was confirmed by observing overlap of

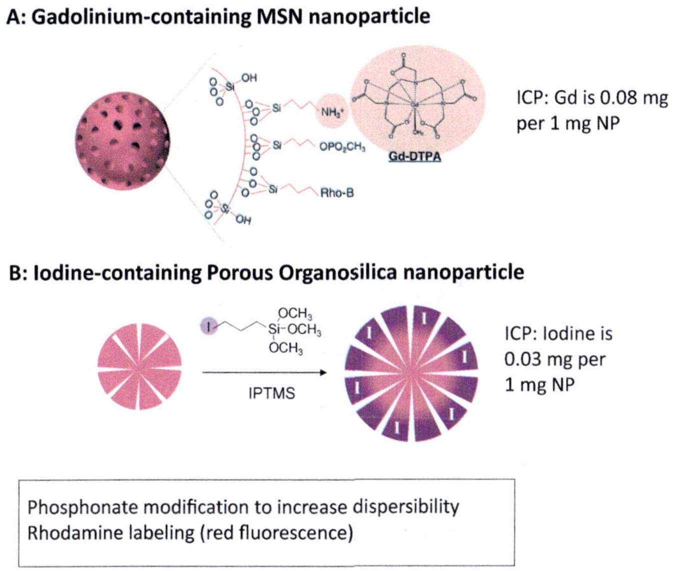

Fig. 4 Nanoparticles containing high Z elements. (A) Gadolinium containing MSN nanoparticles were synthesized by surface modifying with amine groups and then adding gadopentetic acid. (B) Iodine containing porous organosilica nanoparticles were synthesized using an iodine-containing precursor IPTMS. The amount of high Z element in the nanoparticle was measured by ICP. The nanoparticles were modified with phosphonate to increase dispersibility and were labeled with Rhodamine-B.

green fluorescence of GFP-expressing ovarian cancer cells with red fluorescence of nanoparticles using confocal microscopy.

Irradiation of tumor spheroids loaded with Gd-MSN or IPO was carried out at the SPring-8 synchrotron located in Harima, Japan. Irradiation was carried out for up to 1 h and then the spheroids were incubated for 2–3 days. By the γH2AX assay, it was found that double-strand breaks were generated by the irradiation that was dependent on the presence of high Z element. During the 3-day incubation, apoptosis was induced as detected by the TUNEL assay. At the end of the incubation, the spheroids were completely destructed (Fig. 6).

The above results suggest an approach to effective Auger therapy. In this approach, nanoparticles are used to deliver high Z atoms to tumor. Passive tumor accumulation due to leaky vasculature (enhanced permeability retention, EPR effect) as well as active targeting using for example folate that binds to folate receptor that is overexpressed in cancer cells can be used. After the nanoparticles are accumulated in the tumor, irradiation with monochromatic X-ray will be carried out.

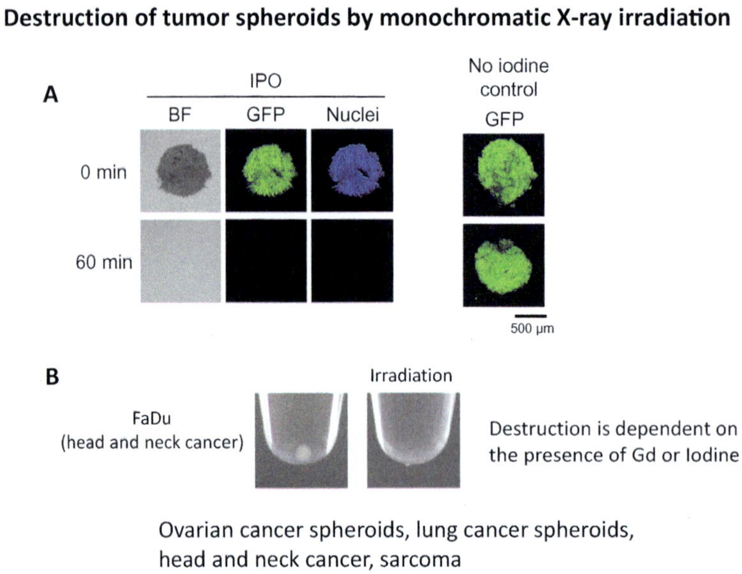

Fig. 5 Destruction of tumor spheroids by monochromatic X-ray irradiation. (A) Monochromatic X-ray irradiation of tumor spheroids (prepared with GFP-expressing ovarian cancer cells) loaded with nanoparticles (IPO) resulted in complete destruction of spheroids as detected by bright field (BF), green fluorescence of GFP or blue fluorescence of nuclear dye. (B) The destruction of head and neck cancer spheroid can be viewed by observing tumor spheroids. The destruction was dependent on the presence of high Z element. Similar results were obtained with lung cancer spheroids as well as with sarcoma spheroids. *Microscope photos are taken from the results published in reference Y. Higashi, K. Matsumoto, H. Saitoh, A. Shiro, Y. Ma, M. Laird, et al. Iodine containing porous organosilica nanoparticles trigger tumor spheroids destruction upon monochromatic X-ray irradiation: DNA breaks and K-edge energy X-ray. Sci. Rep. 11 (2021) 14192, but the overall figure is significantly changed.*

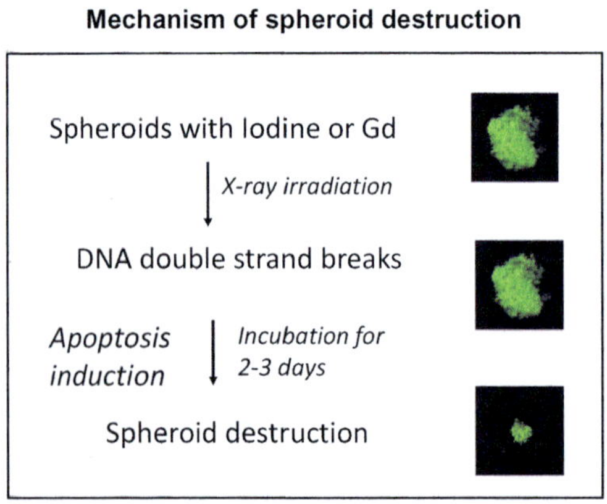

Fig. 6 Mechanism of spheroid destruction. DNA double-strand breaks were formed after monochromatic X-ray irradiation. During the incubation in CO_2 incubator for 2–3 days, apoptosis induction occurs that could be detected by the use of TUNEL assay.

3.4 Comparison of Auger therapy with conventional X-ray therapy

A comparison of some features of conventional X-ray therapy with those of the Auger therapy using high Z elements is described in Fig. 7. The main difference between the conventional X-ray and the Auger therapy is that, while the conventional therapy uses X-ray as the only player in the therapy, the Auger therapy uses not only X-ray but also high Z elements such as gadolinium and iodine. In this sense, the Auger therapy is a binary approach (two components) that uses X-ray and high Z elements that will be irradiated with the X-ray to generate electrons. It is the electron that damages DNA.

Conventional X-ray therapy uses white X-ray which is a mixture of X-ray waves having a variety of energy levels. To increase penetration, high energy X-ray has been increasingly used. In the case of the Auger therapy, we use monochromatic X-ray that has a single energy level. Because monochromatic X-rays interact strongly with high Z elements but interact much less with other elements, it is expected that potential side effect will be much less when monochromatic X-rays are used.

Low oxygen conditions occur inside human body. In particular, low oxygen environment is detected at the center of the tumor and this contributes to the resistance to radiation therapy. Under low oxygen conditions, generation of reactive oxygen species is suppressed and this decreases indirect effect of radiation on DNA. In contrast, the Auger therapy using

Conventional radiation therapy	**Auger therapy using high Z element**
X-ray is the main player	X-ray and high Z elements such as gadolinium and iodine
White X-ray	**Monochromatic X-ray**
High energy X-ray is often used	**Low energy X-ray** with optimum energy is used
Side effects	Side effects may be less
Does not work well in low oxygen conditions	Maybe, less affected by hypoxic conditions

Fig. 7 Comparison of Auger therapy with conventional X-ray therapy. Features of the Auger therapy using high Z elements are compared with the features of conventional X-ray therapy.

high Z element has the ability to generate electrons very close to DNA. This favors direct effect on DNA rather than indirect effect on DNA.

Acknowledgment

This work was supported by JSPS KAKENHI Grant Nos. JP20H00331 and JP21K14502.

References

[1] F. Bray, et al., WHO international agency for research on cancer, CA Cancer J. Clin. 68 (2020) 394.
[2] J. Kleeff, M. Korc, M. Apte, et al., Pancreatic cancer, Nat. Rev. Dis. Primers 2 (2016) 1–22.
[3] F.G. Lawton, E.J. Pavlik, Perspectives on ovarian cancer 1809 to 2022 and beyond, Diagnostics 12 (2022) 791.
[4] M. Yashiro, T. Matsuoka, Scirrhous gastric cancer: a critical review, OA Cancer 2 (2014) 4.
[5] F. Drablos, E. Feyzi, et al., Alkylating damage in DNA and RNA, DNA Repair 3 (2004) 1389.
[6] D. Fu, J. Calvo, L. Samson, SERIES: genomic instability in cancer balancing repair and tolerance of DNA damage caused by alkylating agents, Nat. Rev. Cancer 12 (2012) 104.
[7] K. Wang, J. Lu, R. Li, The events that occur when cisplatin encounters cells, Coord. Chem. Rev. 151 (1996) 53–88.
[8] L. Kelland, The resurgence of platinum-based cancer therapy, Nat. Rev. Cancer 7 (2007) 573–584.
[9] T. Kishimoto, et al., Different effects of cisplatin and transplatin on the higher-order structure of DNA and gene expression, Int. J. Mol. Sci. 21 (34) (2020) 1–16, https://doi.org/10.3390/ijms21010034.
[10] W.B. Parker, Enzymology of purine and pyrimidine antimetabolites used in the treatment of cancer, Chem. Rev. 109 (2009) 2880.
[11] D.B. Longley, et al., 5-fluorouracil: mechanisms of action and clinical strategies, Nat. Rev. Cancer 3 (2003) 330.
[12] S. Noble, K.L.G. Goa, A review of its pharmacology and clinical potential in non-small cell lung cancer and pancreatic cancer, Drugs 54 (1997) 447–472.
[13] S. Cutts, A. Nudelman, et al., The power and potential of doxorubicin-DNA adducts, IUBMB Life 57 (2005) 73–81.
[14] Y. Pommier, E. Leo, H.L. Zhang, C. Marchand, DNA topoisomerases and their poisoning by anticancer and antibacterial drugs, Chem. Biol. 17 (2010) 421.
[15] M. Sastry, R. Fiala, R. Lipman, M. Tomasz, D.J. Patel, Solution structure of the mono-alkylated mitomycin C-DNA complex, J. Mol. Biol. 247 (1995) 338.
[16] L. Povirk, DNA damage and mutagenesis by radiomimetic DNA-cleaving agents: bleomycin, neocarzinostatin and other endiynes, Mutat. Res. 355 (1996) 71–89.
[17] J.J. Champoux, DNA topoisomerases: structure, function, and mechanism, Annu. Rev. Biochem. 70 (2001) 369.
[18] S.J. Froelich-Ammon, Topoisomerase poisons: harnessing the dark side of enzyme mechanism, J. Biol. Chem. 270 (1995) 214295.
[19] C. Garibaldi, B. Jereczek-Fossa, G. Marvaso, et al., Recent advances in radiation oncology, eCancer 11 (2017) 785.
[20] R. Mohan, A review of proton therapy—current status and future directions, Prec. Radiat. Oncol. 6 (2022) 164–176, https://doi.org/10.1002/pro6.1149.
[21] T.D. Malouff, A. Mahajan, S. Krishnan, et al., Carbon ion therapy: a modern review of an emerging technology, Front. Oncol. 10 (2020) 82.

[22] P. Auger, The auger effect, Surf. Sci. 48 (1975) 1–8.
[23] R.F. Martin, L.E. Feinendegen, The quest to exploit the Auger effect in cancer radiotherapy—a reflective review, Int. J. Radiat. Biol. 92 (2016) 617–632.
[24] K. Matsumoto, H. Saitoh, T.L.H. Doan, A. Shiro, K. Nakai, A. Komatsu, et al., Destruction of tumor mass by gadolinium-loaded nanoparticles irradiated with monochromatic X-rays: implications for the Auger therapy, Sci. Rep. 9 (2019) 1–10.
[25] Y. Higashi, K. Matsumoto, H. Saitoh, A. Shiro, Y. Ma, M. Laird, et al., Iodine containing porous organosilica nanoparticles trigger tumor spheroids destruction upon monochromatic X-ray irradiation: DNA breaks and K-edge energy X-ray, Sci. Rep. 11 (2021) 14192.

CHAPTER THREE

The enzymes first edition

Fuyuhiko Tamanoi[a,b,*]
[a]Institute for Integrated Cell-Material Sciences, Institute for Advanced Study, Kyoto University, Kyoto, Japan
[b]Dept. of Microbiology, Immunology & Molecular Genetics, University of California, Los Angeles, CA, United States
*Corresponding author: e-mail address: tamanoi.fuyuhiko.2c@kyoto-u.ac.jp

Contents

Abstract

The Enzymes series was initiated in 1950 with the publication of a book entitled, "The Enzymes: Chemistry and Mechanism of Action" edited by James B. Sumner and Karl Myerback. There are two parts, Part 1 and Part 2 and the book contains 78 chapters. Authors and chapter titles for Part 1 and Part 2 are listed.

Overview

In 1950, The Enzymes series was initiated by James B. Sumner and Karl Myrback. In the Foreword for the first volume, Sumner and Myrback wrote:

"Recent years have seen a remarkable advance in every phase of enzymology. Indeed, research in this field has progressed to such an extent that a survey of its findings in comprehensive form has become increasingly imperative. Enzymology has become the central province of biochemistry. Enzymes are the principal tools of the living cell; non-enzymatic substances and conditions known to be important for the functioning of living matter in an ever-increasing number of cases have been shown to exert their action by means of enzymes, sometimes as components of enzyme systems, and in other instances by regulating the rate or mode of action of an existent enzyme. It is therefore self-evident that enzymology must form an essential part of all biological sciences. Equally, the problem of the chemical nature of enzymes and its relation to their catalytic activity is one of the most intriguing questions of organic and physical chemistry. … It is our aim, therefore,

The Enzymes, Volume 52
ISSN 1874-6047
https://doi.org/10.1016/bs.enz.2022.10.003

to present systematically the accumulated knowledge in the various phases of enzymology as a comprehensive survey which will be the most efficient service both to those already working in the field and to those preparing to enter it. . . ."

CONTENTS PART 1

8. **Enzymes and Immunology**
J.R. MARRACK
I. Introduction
II. Reactions between Antibodies and Antigens
III. Specific Combination of Antigen and Antibody
IV. Effects of Disorganization of Molecules of Antigens and Antibodies
V. Enzymes as Antigens
VI. Toxic Enzymes
VII. Complement.
9. **Enzymes Hydrolyzing Fats and Esters**
ROBERT AMMON AND MAIRE JAARMA
I. Classification of the Esterases and Remarks on the Organization of the Field
II. Brief Survey of Methods of Determination and of Esterase Preparations
III. The Individual Esterases
10. **Acetylcholine Esterase and Cholinesterase**
KLAS-BERTIL AUGUSTINSSON
I. Introduction
II. Existence of Specific Esterases for Acetylcholine and Other Choline Esters
III. Measurement of Activity
IV. Occurrence of Acetylcholine Esterase and Cholinesterase
V. Enzyme Preparations
VI. Activators and Nature of the Active Groups
VII. Kinetics
VIII. Inhibitors
IX. Physiological Significance of Acetylcholine-Hydrolyzing Enzymes
11. **Phosphatases**
JEAN ROCHE
I. Historical Introduction
II. General Biochemistry of the Phosphatases
III. Biochemistry of the Various Types of Phosphatases
IV. Physiological Chemistry of the Phosphatases
V. Pathological Chemistry of the Phosphatases
VI. General Conclusions

CONTRIBUTORS to VOLUME I, PART 1

ROBERT AMMON, *Chemisch-Pharmazeutische Fabrik, Illertissen, Bavaria, Germany.*

KLAS-BERTIL AUGUSTINSSON, *Biochemical Institute, University of Stockholm, Stockholm, Sweden.*

A. L. DOUNCE, *Biochemistry Department, Strong Memorial Hospital, Rochester, New York.*

WILLIAM H. FISHMAN, *Tufts College Medical School, Boston, Massachusetts.*

CLAUDE FROMAGEOT, *Laboratoire de Chimie biologique, Faculte des Sciences, Paris, France.*

J. S. GOTS, *University of Pennsylvania School of Medicine, Philadelphia, Pennsylvania.*

ALFRED GOTTSCHALK, *The Walter and Eliza Hall Institute of Medical Research, Melbourne, Australia.*
BURCKHARDT HELFERICH, *University of Bonn, Germany.*
BJORN INGELMAN, *Institute of Biochemistry and Physical Chemistry, University of Uppsala, Uppsala, Sweden.*
MAIRE JAARMA, *Biokemiska Institutet, Stockholms, Hogskola, Stockholm, Sweden.*
INES MANDL, *Interchemical Corporation, New York, New York.*
J. R. MARRACK, *London Hospital, Medical College, London, England.*
L. MASSART, *Biochemical Institute of the University, Ghent, Belgium.*
E. A. MOELWYN-HUGHES, *Department of Physical Chemistry, University of Cambridge, England.*
KARL MYRBACK, *Institute for Organic Chemistry and Biochemistry, University of Stockholm, Stockholm, Sweden.*
CARL NEUBERG, *Polytechnic Institute of Brooklyn, Brooklyn, New York.*
GUNNAR NEUMULLER, *Institute for Organic Chemistry and Biochemistry, University of Stockholm, Stockholm, Sweden.*
JEAN ROCHE, *College de France, Paris, France.*
M. G. SEVAG, *University of Pennsylvania School of Medicine, Philadelphia, Pennsylvania.*
S. SPIEGELMAN, *Department of Bacteriology, University of Illinois, Urbana, Illinois.*
E. STEERS, *University of Pennsylvania School of Medicine, Philadelphia, Pennsylvania.*
JAMES B. SUMNER, *Laboratory of Enzyme Chemistry, Cornell University, Ithaca, New York.*
STIG VEIBEL, *Department of Organic Chemistry of the University, Copenhagen, Denmark.*

CONTENTS OF VOLUME I, PART 2

27. Allantoinase and Allantoicase. By M. LASKOWSKI
28. Hippuricase (Histozyme). By F. LEUTHARDT
29. Nucleolytic Enzymes. By M. LASKOWSKI
30. Enzymes as Essential Compounds of Bacterial and Animal Toxins. By E.A. ZELLER
31. Phosphorylases-Phosphorolysis and Synthesis of Saccharides. By W.Z. HASSID, M. DOUDOROFF, and H.A. BARKER
32. Transaminases. By PHILIP P. COHEN
33. Transmethylases. By THEODORE L. SOURKES
34. Enzymes and Coagulation. I. Milk Coagulation By N.J. BERRIDGE
35. Blood coagulation. By WALTER H. SEEGERS
36. Histidase and Urocanase. By F. LEUTHARDT
37. Penicillinase. By E.P. ABRAHAM
38. Thiaminase. By ROBERT S. HARRIS
39. Enolase. By OTTO MEYERHOF
40. Fumarase and Aconitase. By SEVERO OCHOA
41. Desulfhydrases. By CLAUDE FROMAGEOT
42. Aspartase. By JORMA ERKAMA and ARTTURI I. VIRTANEN
43. Carbonic Anhydrase. By F.J.W. ROUGHTON and A.M. CLARK
44. Theory of Oxidation-Reduction. By LEONOR MICHAELIS
45. Thermodynamics and Mechanism of the Phosphate Bond. By NATHAN O. KAPLAN
46. Transphosphorylating Enzymes of Fermentation. By SIDNEY P. COLOWICK
47. Dephosphorylation of Adenosine Triphosphate. By H.M. KALCKAR
48. Aldolase and lsomerase. By OTTO MEYERHOF
49. Keto Acid Decarhoxylases. By BIRGIT VENNESLAND
50. Amino Acid Decarboxylases. By OTTO SCHALES
51. Desulfinase. By CLAUDE FROMAGEOT
52. Codehydrogenases I and II and Apcenzymes. By F. SCHLENK
53. Succinic Dehydrogenase. By F. SCHLENK
54. Hydrogenase and Hydrogenylase. By W.W. UMBREIT
55. Flavin-Containing Enzymes. By H. THEORELL
56. Iron-Containing Enzymes. By BRITTON CHANCE, KARL-GUSTAV PAUL, and H. THECRELL
57. The Copper Oxidases. By C.R. DAWSON and W.B. TARPLEY
58. Oxidation of Amino Acids. By H.A. KREBS

59. Oxidation of Amines. By E.A. ZELLER
60. Lipoxidase or Unsaturated-Fat Oxidase. By RALPH T. HOLMAN and SUNE BERGSTROM
61. Enzymes in Luminescence. By E. NEWTON HARVEY
62. Oxidation of Organic Sulfur. By CLAUDE FROMAGEOT
63. Anaërobic Glycolysis, Respiration, and the Pasteur Effect. By FRANK DICKENS
64. Yeast and Mold Fermentations. By F.F. NORD and SIDNEY WEISS
65. Fermentations by Bacteria. By S.R. ELSDEN
66. Enzymes in Detoxication. By F. BERNHEIM
67. Urea Synthesis. By H.A. KREBS
68. Peptide Bond Synthesis. By PHILLIP P. COHEN
69. Enzymatic Synthesis of Glutamine. By PHILLIP P. COHEN
70. Synthesis of Acetylcholine. Choline Acetylase. By KLAUS-BERTIL AUGUSTINSSON
71. Oxidative Assimilation by Microorganisms. By C.E. CLIFTON
72. Carbonic Acid Fixation. By SEVERO OCHOA
73. Carbohydrate-Fat Conversion. By F.L. BREUSCH
74. Photosynthesis in Green Plants. By WALTER E. LOOMIS
75. Bacterial Photosynthesis. By C.B. VAN NIEL
76. Nitrogen Assimilation. By ARTTURI I. VIRTANEN and NILLO RAUTANEN
77. Tumor Enzymology. By JESSE P. GREENSTEIN and ALTON MEISTER
78. Enzyme Technology. By A.F. LANGLYKKE, C.V. SMYTHE, and D. PERLMAN

CHAPTER FOUR

The enzymes second edition part 1

Fuyuhiko Tamanoi[a,b,*]
[a]Institute for Integrated Cell-Material Sciences, Institute for Advanced Study, Kyoto University, Kyoto, Japan
[b]Dept. of Microbiology, Immunology & Molecular Genetics, University of California, Los Angeles, CA, United States
*Corresponding author: e-mail address: tamanoi.fuyuhiko.2c@kyoto-u.ac.jp

Contents

Abstract

Publication of the second edition of The Enzymes series was initiated in 1959 and eight volumes were published. This chapter describes volumes 1–3. All the eight volumes were edited by Paul D. Boyer, Henry Lardy and Karl Myerback. Authors and chapter titles are listed.

Overview

In 1958, Paul Boyer, Henry Lardy and Karl Myerback initiated compiling second edition of the Enzymes. Eight volumes were published and this chapter provides information on volumes 1–3 that encompass 35 chapters.

In the Preface, the editors wrote:

"The over-all aim of the treatise will be to present fully the available information about enzymes and enzyme action on a molecular level. Our knowledge of enzymes has progressed sufficiently to make organization of the treatise at such a level possible, and indeed, essential, for a fuller understanding of the field. To achieve the desired end, complete revision of the previous edition appears necessary. Much of the material included in the contemplated edition will be entirely new or supplemental to that of the previous edition, and thus the previous edition will continue to serve as a useful reference work in certain areas. Also, the availability of the companion treatise, "Methods in Enzymology" makes unnecessary the inclusion of details on methods in "The Enzymes," and thus allows a fuller discussion of properties, mechanism, and chemistry of the enzymes themselves."

The Enzymes, Volume 52
ISSN 1874-6047
https://doi.org/10.1016/bs.enz.2022.10.004

SECOND EDITION

CONTRIBUTORS TO VOLUME 1

ROBERT A. ALBERTY, *Department of Chemistry, University of Wisconsin, Madison, Wisconsin*

S.A. BERNHARD, *Naval Medical Research Institute, Bethesda, Maryland*

D.J. BOTTS, *Department of Biochemistry, Dartmouth Medical School, Hanover, New Hampshire*

P.D. BOYER, *Department of Physiological Chemistry, University of Minnesota, Minneapolis, Minnesota*

H. FRAENKEL-CONRAT, *Department of Virology, University of California, Berkeley, California*

S.L. FRIESS, *Naval Medical Research Institute, Bethesda, Maryland*

PHILIP GEORGE, John Harrison Laboratory of Chemistry, University of Pennsylvania, Philadelphia, Pennsylvania

J.S. GRIFFITH, *Department of Theoretical Chemistry, The University of Cambridge, Cambridge, England*

H. GUTFREUND, *National Institute for Research in Dairying, Shinfield, near Reading, England*

JOHN Z. HEARON, *National Institutes of Health, Bethesda, Maryland*

D.E. KOSHLAND, JR., *Brookhaven National Laboratory, Upton, New York*

K.U. LINDERSTROM-LANG, *Carlsberg Laborator y, Copenhagen, Denmark*

RUFUS LUMRY, *School of Chemistry, University of Minnesota, Minneapolis, Minnesota*

M.F. MORALES, *Department of Biochemistry, Dartmouth Medical School, Hanover, New Hampshire*

ARTHUR B. PARDEE, *Virus Laboratory and Department of Biochemistry, University of California, Berkeley, California*

M.R. POLLOCK, *National Institute for Medical Research, London, England*

J.A. SCHELLMAN, *School of Chemistry, University of Minnesota, Minneapolis, Minnesota*

HAROLD L. SEGAL, *Biochemistry Department, University of Pittsburgh, School of Medicine, Pittsburgh, Pennsylvania*

F.H. WESTHEIMER, *Department of Chemistry, Harvard University, Cambridge, Massachusetts*

R.J.P. WILLIAMS, *Inorganic Chemistry Laboratory and Wadham College, Oxford, England*

CONTENTS

VOLUME 1

AUTHOR INDEX
SUBJECT INDEX

VOLUME 2

PURINE AND PYRIMIDINE NUCLEOTIDES AND PHOSPHAGENS

CONTRIBUTORS TO VOLUME 2

HELMUT BEINERT, *Institute for Enzyme Research, University of Wisconsin, Madison, Wisconsin*

ROBERT M. BOCK, *Department of Biochemistry, University of Wisconsin, Madison, Wisconsin*

ALEXANDER E. BRAUNSTEIN, *The USSR Academy of Medical Sciences, Moscow, USSR*

CARLOS E. CARDINI, *Instituto de Investigaciones Bioquimicas, Fundaci6n Campomar, and Facultad de Ciencias Exactas y Naturales, Buenos Aires, Argentina*

A. H. ENNOR, *Department of Biochemistry, The John Curtin School of Medical Research, Australian National University, Canberra, Australia*

EUGENE P. KENNEDY, *Department of Biochemistry, University of Chicago, Chicago, Illinois*

W. EUGENE KNOX, *New England Deaconess Hospital, Boston, Massachusetts*

LUIS F. LELOIR, *Instituto de Investigaciones Bioquimicas, Fundacion Campomar, and Facultad de Ciencias Exactas y Naturales, Buenos Aires, Argentina*

DAVID E. METZLER, *Department of Chemistry, Iowa State College, Ames, Iowa*

J. F. MORRISON, *Department of Biochemistry, The John Curtin School of Medical Research, Australian National University, Canberra, Australia*

JESSE C. RABINOWITZ, *Department of Biochemistry, University of California, Berkeley, California*

MERTON F. UTTER, *Western Reserve University, Cleveland, Ohio*

CONTENTS OF VOLUME 2

CONTRIBUTORS TO VOLUME 2

PREFACE CONTENTS OF VOLUME 1; TENTATIVE CONTENTS OF VOLUMES 3, 4, and 5

Purine and Pyrimidine Nucleotides and Phosphagens

VOLUME 3
PTOSTHETIC GROUPS AND COFACTORS (PART B)
CONTRIBUTORS TO VOLUME 3

GILBERT ASHWELL, *Laboratory of Chemical Pharmacology, National Heart Institute and the Laboratory of Biochemistry and Metabolism, National Institute of Arthritis and Metabolic Diseases, National Institutes of Health, Bethesda, Maryland*
P.D. BOYER, *Department of Physiological Chemistry, University of Minnesota, Minneapolis, Minnesota*
J.J. BURNS, *Laboratory of Chemical Pharmacology, National Heart Institute and the Laboratory of Biochemistry and Metabolism, National Institute of Arthritis and Metabolic Diseases, National Institutes of Health, Bethesda, Maryland*
HENRIK DAM, *Danmarks Tekniske H¢jskole, Copenhagen, Denmark*
RUTH HUBBARD, *Biological Laboratories at Harvard University, Cambridge, Massachusetts*
L. JAENICKE, *Max-Planck-Institut fur Zellchemie, Munchen, Germany*
MARVIN J. JOHNSON, *Department of Biochemistry, University of Wisconsin, Madison, Wisconsin*
NATHAN O. KAPLAN, *Graduate Department of Biochemistry, Brandeis University, Waltham, Massachusetts*
EDWARD M. KOSOWER, *Department of Chemistry, University of Wisconsin, Madison, Wisconsin*
F. LYNEN, *Max-Planck-Institut fur Zellchemie, Munchen, Germany*
K.G. PAUL, *Nobel Medical Institute, Biochemical Department, Stockholm, Sweden*
LESTER J. REED, *Clayton Foundation Biochemical Institute and Department of Chemistry, University of Texas, Austin, Texas*
EBBE S¢NDERGAARD, *Danmarks Tekniske H¢jskole, Copenhagen, Denmark*
BERT L. VALLEE, *Biophysics Research Laboratory, Harvard Medical School and Peter Bent Brigham Hospital, Boston, Massachusetts*
GEORGE WALD, *Biological Laboratories at Harvard University, Cambridge, Massachusetts*

CONTENTS

CHAPTER FIVE

The enzymes second edition part 2

Fuyuhiko Tamanoi[a,b,*]
[a]Institute for Integrated Cell-Material Sciences, Institute for Advanced Study, Kyoto University, Kyoto, Japan
[b]Dept. of Microbiology, Immunology & Molecular Genetics, University of California, Los Angeles, CA, United States
*Corresponding author: e-mail address: tamanoi.fuyuhiko.2c@kyoto-u.ac.jp

Contents

Abstract

Publication of the second edition of The Enzymes series was initiated in 1959 and eight volumes were published. This chapter describes volumes 4–8. All eight volumes were edited by Paul D. Boyer, Henry Lardy and Karl Myerback. Authors and chapter titles are listed.

Overview

This chapter provides information on volumes 4–8 of the second edition put together by Paul Boyer, Henry Lardy and Karl Myerback. The number of chapters in each volume is as follows.

Volume 4: 33 chapters
Volume 5: 34 chapters
Volume 6: 31 chapters
Volume 7: 26 chapters
Volume 8: 14 chapters

SECOND EDITION
VOLUME 4
CONTRIBUTORS TO VOLUME 4

FRANK B. ABLONDI, *Biochemistry Department, Biochemistry Research Section, Lederle Laboratories Division, American Cyanamid Company, Pearl River, New York*

KLAS-BERTIL AUGUSTINSSON, *Institute of Organic Chemistry and Biochemistry, University of Stockholm, Stockholm, Sweden*

The Enzymes, Volume 52
ISSN 1874-6047
https://doi.org/10.1016/bs.enz.2022.10.005

D. JOE BAUGHMAN, *Massachusetts Institute of Technology, Cambridge, Massachusetts*

F.A. BOVEY, *Central Research Laboratories, Minnesota Mining and Manufacturing Company, St. Paul, Minnesota*

P. DESNUELLE, *Laboratoire de Chimie Biologique, Faculte des Sciences, Marseille, France*

EDMOND H. FISCHER, *Department of Biochemistry, University of Washington, Seattle, Washington*

DEXTER FRENCH, *Department of Biochemistry and Biophysics, Iowa State University, Ames, Iowa*

JOSEPH S. FRUTON, *Department of Biochemistry, Yale University, New Haven, Connecticut*

ALFRED GOTTSCHALK, *Department of Microbiology, Australian National University, Canberra, Australia*

DAVID M. GREENBERG, *Department of Biochemistry, University of California School of Medicine, San Francisco, California*

JAMES J. HAGAN, *Biochemistry Department, Biochemistry Research Section, Lederle Laboratories Division, American Cyanamid Company, Pearl River, New York*

BUNJI HAGIHARA, *Institute for Enzyme Research, University of Wisconsin, Madison, Wisconsin; Faculty of Science, Osaka University, Osaka, Japan*

ROBERT L. HILL, *Departments of Biochemistry and Medicine, University of Utah College of Medicine, Salt Lak e City, Utah*

PHILIP HOFFMAN, *Department of Medicine, College of Physicians and Surgeons, Columbia University, and the Edward Daniels Faulkner Arthritis Clinic, Presbyterian Hospital, New York, New York*

B.H.J. HOFSTEE, *Palo Alto Medical Research Foundation, Palo Alto, California*

PIERRE JOLLES, *Laboratoire de Chimie Biologique, Universite de Paris, Paris, France*

J.R. KIMMEL, *Departments of Biological Chemistry and Medicine, University of Utah College of Medicine, Salt Lake City, Utah*

J. LARNER, *Department of Pharmacology, Western Reserve University School of Medicine, Cleveland, Ohio*

YA PIN LEE, *Institute for Enzyme Research, University of Wisconsin, Madison, Wisconsin*

G.A. LEVVY, *Department of Enzymology, Rowett Research Institute, Bucksburn, Aberdeen, Scotland*

ALFRED LINKER, *Department of Medicine, College of Physicians and Surgeons, Columbia University, and the Edward Daniels Faulkner Arthritis Clinic, Presbyterian Hospital, New York, New York*

OM PRAKASH MALHOTRA, *Chemisches Laboratorium der Universitat, Freiburg, Germany*

C.A. MARSH, *Department of Enzymology, Rowett Research Institute, Bucksburn, Aberdeen, Scotland*

KARL MEYER, *Department of Medicine, College of Physicians and Surgeons, Columbia University, and the Edward Daniels Faulkner Arthritis Clinic, Presbyterian Hospital, New York, New York*

KENT D. MILLER, *Division of Laboratories and Research, New York State Department of Health, Albany, New York*

L.A. MOUNTER, *Department of Biophysics, Medical College of Virginia, Richmond, Virginia*

D.K. MYERS, *Biology Branch, Atomic Energy of Canada Limited, Chalk River, Ontario, Canada*

KARL MYRBACK, *Institute for Organic Chemistry and Biochemistry, University of Stockholm, Stockholm, Sweden*

HANS NEURATH, *Department of Biochemistry, University of Washington, Seattle, Washington*

M.R. POLLOCK, *Bacterial Physiology Division, National Institute for Medical Research, London, England*

EUGENE ROBERTS, *Department of Biochemistry, Medical Research Institute, City of Hope Medical Center, Duarte, California*

KARL H. SLOTTA, *Departments of Biochemistry and of Medicine, University of Miami Medical School, Miami, Florida*

EMIL L. SMITH, *Laboratory for the Study of Hereditary and Metabolic Disorders, and the Departments of Biochemistry and Medicine, University of Utah College of Medicine, Salt Lake City, Utah*

Eric A. STEIN, *Department of Biochemistry, University of Washington School of Medicine, Seattle, Washington*

J.E. VARNER, *Department of Agricultural Biochemistry, Ohio State University, Columbus, Ohio*

KURT WALLENFELS, *Chemisches Laboratorium der Universitat, Freiburg, Germany*

DAVID F. WAUGH, *Department of Biology, Massachusetts Institute of Technology, Cambridge, Massachusetts*

IRWIN B. WILSON, *College of Physicians and Surgeons, Columbia University, New York, New York*

SAM S. YANARI, *Central Research Laboratories, Minnesota Mining and Manufacturing Company, St. Paul, Minnesota*

CONTENTS

CONTRIBUTORS TO VOLUME 5

ROBERT A. ALBERTY, *Chemistry Department, University of Wisconsin, Madison, Wisconsin*

CHRISTIAN B. ANFINSEN, *Laboratory of Cellular Physiology and Metabolism, National Heart Institute, National Institutes of Health, Public Health Service, U. S. Department of Health, Education and Welfare, Bethesda, Maryland*

DAYID H. BROWN, *Department of Biological Chemistry, Washington University School of Medicine, St. Louis, Missouri*

WILLIAM L. BYRNE, *Department of Biochemistry, Duke University Medical Center, Durham, North Carolina*

MILDRED COHN, *Department of Biological Chemistry, Washington University School of Medicine, St. Louis, Missouri*

CARL F. CORI, *Department of Biological Chemistry, Washington University School of Medicine, St. Louis, Missouri*

ROBERT P. DAVIS, *Department of Medicine, Albert Einstein College of Medicine, Yeshiva University, New York, New York*

SHERMAN R. DICKMAN, *Department of Biological Chemistry, University of Utah, College of Medicine, Salt Lak e City, Utah*

M. DOUDOROFF, *Department of Bacteriology, University of California, Berkeley, California*

MORRIS FRIEDKIN, *Department of Pharmacology, Tufts University School of Medicine, Boston, Massachusetts*

DAVID M. GREENBERG, *Department of Biochemistry, University of California School of Medicine, San Francisco, California*

M. GRUNBERG-MANAGO, *Department of Biochemistry, Institut de Biologie Physico-Chimique, Paris, France*

PHILIP HANDLER, *Department of Biochemistry, Duke University School of Medicine, Durham, North Carolina*

G. HARTMANN, *Chemisches Laboratorium der Universitat, Institut fur Biochemie, und Max-Planck-Institut fur Zellchemie, Munchen, Germany*

LEON A. HEPPEL, *National Institute of Arthritis and Metabolic Diseases, National Institutes of Health, Public Health Service, U. S. Department of Health, Education and Welfare, Bethesda, Maryland*

JOHN lMSANDE, *Department of Biochemistry, Duke University School of Medicine, Durham, North Carolina*

HERMAN KALCKAR, *Department of Biology and McCollum-Pratt Institute, The Johns Hopkins University, Baltimore, Maryland*

H. GOBIND KHORANA, *British Columbia Research Council, University of British Columbia, Vancouver, British Columbia*

W. WAYNE KIELLEY, *Laboratory of Cellular Physiology and Metabolism, National Heart Institute, National Institutes of Health, Public Health Service, U. S. Department of Health, Education and Welfare, Bethesda, Maryland*

MOSES KUNITZ, *The Rockefeller Institute, New York, New York*

M. LASKOWSKI, SR., *Biochemical Laboratory for Cancer Research, Marquette University School of Medicine, Milwaukee, Wisconsin*

F. LYNEN, *Chemisches Laboratorium der Universitat, Institut fur Biochemie, und Max-Planck-Institut fur Zellchemie, Munchen, Germany*

Bo G. MALMSTROM, *Institute of Biochemistry, University of Uppsala, Uppsala, Sweden*

E.S. MAXWELL, *National Institute of Arthritis and Metabolic Diseases, National Institutes of Health, Public Health Service, U. S. Department of Health, Education and Welfare, Bethesda, Maryland*

JAMES ALLEN OLSON, *Department of Biochemistry, College of Medicine, University of Florida, Gainesville, Florida*

E. RACKER, *Division of Nutrition and Physiology, The Public Health Research Institute of the City of New York, New York*

PHILLIPS W. ROBBINS, *The Rockefeller Institute, New York, New York*

WILLIAM J. RUTTER, *Division of Biochemistry, Department of Chemistry, University of Illinois, Urbana, Illinois*

GERHARD SCHMIDT, *Boston Dispensary and Department of Biochemistry, Tufts University School of Medicine, Boston, Massachusetts*

THRESSA C. STADTMAN, *Enzyme Section, National Heart Institute, National Institutes of Health, Public Health Service, U. S. Department of Health, Education and Welfare, Bethesda, Maryland*

JOSEPH R. STERN, *Department of Pharmacology, School of Medicine, Western Reserve University, Cleveland, Ohio*
YALE J. TOPPER, *National Institute of Arthritis and Metabolic Diseases, National Institutes of Health, Public Health Service, U. S. Department of Health, Education and Welfare, Bethesda, Maryland*
MERTON F. UTTER, *Department of Biochemistry, Western Reserve University, Cleveland, Ohio*
FREDERICK H. WHITE, Jr., *Laboratory of Cellular Physiology and Metabolism, National Heart Institute, National Institutes of Health, Public Health Service, U. S. Department of Health, Education and Welfare, Bethesda, Maryland*

VOLUME 5 CONTENTS

CONTRIBUTORS TO VOLUME 6

P.D. BOYER, *Department of Physiological Chemistry, University of Minnesota, Minneapolis, Minnesota.*

RANWELL CAPUTTO, *Biochemistry Section of the Oklahoma Medical Research Institute, Oklahoma City, Oklahoma*

CARLOS E. CARDINI, *Instituto de Investigaciones Bioquimicas "Fundacion Compomar" and Facultad de Ciencias Exactas y Naturales, Buenos Aires, Argentina.*

PHILIP P. COHEN, *Department of Physiological Chemistry, University of Wisconsin, Madison, Wisconsin.*

ROBERT K. CRANE, *Department of Biological Chemistry, Washington University School of Medicine, St. Louis, Missouri.*

LOWELL P. HAGER, *Department of Chemistry, University of Illinois, Urbana, Illinois.*

W.Z. HASSID, *Department of Biochemistry, University of California, Berkeley, California.*

H.G. HERS, *Laboratory of Physiological Chemistry, University of Louvain, Louvain, Belgium.*

WILLIAM P. JENCKS, *Graduate Department of Biochemistry, Brandeis University, Waltham, Massachusetts.*

STEPHEN A. KUBY, *Institute for Enzyme Research, University of Wisconsin, Madison, Wisconsin.*

HENRY LARDY, *Institute for Enzyme Research, University of Wisconsin, Madison, Wisconsin.*
MARTHA LARSSON-RAZNIKIEWICZ, *Institute of Biochemistry, University of Uppsala, Uppsala, Sweden.*
LUIS F. LELOIR, *Instituto de Investigaciones Bioquimicas "Fundacion Campomar" and Facultad de Ciencias Exactas y Naturales, Buenos Aires, Argentina.*
D. McELROY, *McCollum-Pratt Institute and Department of Biology, The Johns Hopkins University, Baltimore, Maryland.*
BO G. MALMSTROM, *Institute of Biochemistry, University of Uppsala, Uppsala, Sweden.*
MARGARET MARSHALL, *Department of Physiological Chemistry, University of Wisconsin, Madison, Wisconsin.*
ALTON MEISTER, *Department of Biochemistry, Tufts University School of Medicine, Boston, Massachusetts.*
VICTOR A. NAJJAR, *Department of Microbiology, Vanderbilt University School of Medicine, Nashville, Tennessee.*
E. F. NEUFELD, *Department of Biochemistry, University of California, Berkeley, California.*
LAFAYETTE NODA, *Department of Biochemistry, Dartmouth Medical School, Hanover, New Hampshire.*
ERNST NOLTMANN, *Institute for Enzyme Research, University of Wisconsin, Madison, Wisconsin.*
ROBERT NORDLIE, *Institute for Enzyme Research, University of Wisconsin, Madison, Wisconsin.*
G. DAVID NOVELLI, *Biology Division, Oak Ridge National Laboratory, Oak Ridge, Tennessee.*
LEWIS I. PIZER, *Department of Biochemistry, University of Pennsylvania, Philadelphia, Pennsylvania.*
MURRAY RABINOWITZ, *Departments of Medicine and Biochemistry, The University of Chicago, Chicago, Illinois.*
S. RATNER, *Division of Nutrition, and Physiology, The Public Health Research Institute of the City of New York, New York.*
PHILLIPS W. ROBBINS, *Massachusetts Institute of Technology, Cambridge, Massachusetts.*
IRWIN A. ROSE, *Department of Biochemistry, Yale University, New Haven, Connecticut.*
M. P. STULBERG, *Biology Division, Oak Ridge National Laboratory, Oak Ridge, Tennessee.*

JOHN VAVRA, *Department of Biological Chemistry, Washington University School of Medicine, St. Louis, Missouri.*
SIDNEY F. VELICK, *Department of Biological Chemistry, Washington University, School of Medicine, St. Louis, Missouri.*
ROBERT H. WEAVER, *Departments of Pediatrics and Physiological Chemistry, University of Wisconsin School of Medicine, Madison, Wisconsin.*

VOLUME 6 CONTENTS

CONTRIBUTORS TO VOLUME 7

AKE AKESON, *Medicinska Nobelinstitutet, Biokemiska Avdelningen Stockholm, Sweden*

TADEUSZ BARANOWSKI, *Department of Biochemistry, Medical School of Wroclaw, Poland*

HELMUT BEINERT, *Institute for Enzyme Research, University of Wisconsin, Madison, Wisconsin*

RONALD BENTLEY, *Department of Biochemistry and Nutrition, Graduate School of Public Health, University of Pittsburgh, Pittsburgh, Pennsylvania*

R. C. BRAY, *Chester Beatty Research Institute, Institute of Cancer Research, Royal Cancer Hospital, Fulham Road, London, England*

ANDERS EHHENBERG, *Medicinska Nobelinstitutet, Biokemiska Avdelningen, Stockholm, Sweden*

CARL FRIEDEN, *Department of Biological Chemistry, Washington University School of Medicine, St. Louis, Missouri*

CHARLES FURFINE, *Department of Biological Chemistry, Washington University School of Medicine, St. Louis, Missouri*

YOUSSEF HATEFI, *Institute for Enzyme Research, University of Wisconsin, Madison, Wisconsin*

WILLIAM B. JAKOBY, *National Institute of Arthritis and Metabolic Diseases, National Institutes of Health, Bethesda, Maryland*

EDNA B. KEARNEY, *Edsel B. Ford Institute for Medical Research, Henry Ford Hospital, Detroit, Michigan*

STEPHEN A. KUBY, *Institute for Enzyme Research, University of Wisconsin, Madison, Wisconsin*

ERNEST KUN, *Departments of Pharmacology and Biochemistry, The University of California School of Medicine, San Francisco, California*

CARL MARTIUS, *Laboratorium fur Biochemie, Eidgenossische Technische Hochschule, Zurich, Switzerland*

VINCENT MASSEY, *Department of Biochemistry, University of Sheffield, Sheffield, England*

ALTON MEISTER, *Department of Biochemistry, Tufts University School of Medicine, Boston, Massachusetts*

ALVIN NASON, *McCollum-Pratt Institute, The Johns Hopkins University, Baltimore, Maryland*

ERNEST A. NOLTMANN, *Institute for Enzyme Research, University of Wisconsin, Madison, Wisconsin*

AGNER P. NYGAARD, *Johan Throne Holst's Institute for Nutrition Research, University of Oslo, Blindern, Oslo, Norway*

REIJI OKAZAKI, *Department of Pharmacology, Washington University School of Medicine, St. Louis, Missouri*

TUNEKO OKAZAKI, *Department of Pharmacology, Washington University School of Medicine, St. Louis, Missouri*

GERHARD W.E. PLAUT, *Laboratory for the Study of Hereditary and Metabolic Disorders and Departments of Biological Chemistry and Medicine, University of Utah College of Medicine, Salt Lake City, Utah*

D. RAO SANADI, *Gerontology Branch, National Heart Institute, National Institutes of Health and the Baltimore City Hospitals, Baltimore, Maryland*

GEORGE W. SCHWERT, *Department of Biochemistry, Medical Center, University of Kentucky, Lexington, Kentucky*

THOMAS P. SINGER, *Edsel B. Ford Institute for Medical Research, Henry Ford Hospital, Detroit, Michigan*

JACK L. STROMINGER, *Department of Pharmacology, Washington University School of Medicine, St. Louis, Missouri*

HORST SUND, *Medicinska Nobelinstitutet, Biokemiska Avdelningen Stockholm, Sweden*

PAUL TALALAY, *Ben May Laboratory for Cancer Research and Biochemistry and Medicine, The University of Chicago, Chicago, Illinois*

HUGO THEORELL, *Medicinska Nobelinstitutet, Biokemiska Avdelningen Stockholm, Sweden*

SIDNEY VELICK, *Department of Biological Chemistry, Washington University School of Medicine, St. Louis, Missouri*

SALIH J. WAKIL, *Department of Biochemistry, Duke University Medical Center, Durham, North Carolina*

DANIEL WELLINER, *Department of Biochemistry, Tufts University School of Medicine, Boston, Massachusetts*

ALFRED D. WINER, *Department of Biochemistry, Medical Center, University of Kentucky, Lexington, Kentucky*

VOLUME 7 CONTENTS

CONTRIBUTORS TO VOLUME 8

H. BLASCHKO, *University Department of Pharmacology, Oxford, England*

CHARLES R. DAWSON, *Department of Chemistry, Columbia University, New York, New York*

DAVID E. GREEN, *Institute for Enzyme Research, University of Wisconsin, Madison, Wisconsin*

ROBERT GOLDBERGER, *Institute for Enzyme Research, University of Wisconsin, Madison, Wisconsin*

OSAMU HAYAISHI, *Department of Medical Chemistry, Kyoto University, Faculty of Medicine, Kyoto, Japan, and Department of Biochemistry, Osaka University, School of Medicine, Osaka, Japan*

SEYMOUR KAUFMAN, *Laboratory of Cellular Pharmacology, National Institute of Mental Health, United States Department of Health, Education and Welfare, United States Public Health Service, National Institutes of Health, Bethesda, Maryland*

H.R. MAHLER, *Indiana University, Bloomington, Indiana*

PETER NICHOLLS, *The Johnson Foundation for Medical Physics, University of Pennsylvania, Philadelphia, Pennsylvania*

S. PALEUS, *Nobel Medical Institute, Stockholm, Sweden*
K. G. PAUL, *Department of Biochemistry, Nobel Medical Institute, Stockholm, Sweden*
G.R. SCHONBAUM, *Department of Biochemistry, University of Alberta., Edmonton, Canada*
GEORGE R. STARK, *Department of Chemistry, Columbia University, New York, New York*
PHILLIPP STRITTMATTER, *Department of Biological Chemistry, Washington University School of Medicine, St. Louis, Missouri*
A. L. TAPPEL, *Department of Food Science and Technology, University of California, Davis, California*
TAKASHI YONETANI, *Johnson Research Foundation, University of Pennsylvania, Philadelphia, Pennsylvania*
E· ALBERT ZELLER, *Department of Biochemistry, Northwestern University Medical School, Chicago, Illinois*

VOLUME 8 CONTENTS

II. Cytochrome b
III. Relationship of Cytochromes b and c_1 to the Structural Protein of the Mitochondrion

4. Mammalian Cytochrome c
S. PALUS AND K.G. PAUL
I. Preparation
II. Properties
III. Modified Cytochrome c
IV. Biosynthesis

5. Microsomal Cytochrome b5 and Cytochrome b5 Reductase
PHILIPP STRITTMATTER
I. Introduction
II. Isolation of Enzymes
III. Molecular Properties
IV. Mechanism of the Reaction

6. Catalases
P. NICHOLLS AND G.R. SCHONBAUM
I. Introduction
II. Physical and Chemical Properties of the Enzyme
III. III.The Enzyme-Substrate Compounds
IV. The Kinetics of the Over-all Reactions
V. Theories of Catalase Action
VI. The Physiological Role of Catalases

7. Peroxidases
K.G. PAUL
I. Introduction
II. Preparative Procedures and Analyses
III. The Primary Substrate (Oxidizing Agent)
IV. The Secondary Substrate ("Oxidizable Compound ")
V. Oxidase-Peroxidase Reactions

8. Lipoxidase
A.L. TAPPEL
I. Occurrence and Function
II. Properties and Specificity
III. Kinetics
IV. Inhibitors
V. Mechanism

9780323993982